Know Your Cattle

Jack Byard

WITHDRAWN

Old Pond Publishing

First published 2008, reprinted 2008

ISBN 978-1-905523-92-4

Published by:
Old Pond Publishing Ltd
Dencora Business Centre
36 White House Road
Ipswich IP1 5LT
United Kingdom

www.oldpond.com

Book design by Liz Whatling
Printed and bound in China

Contents

Acknowledgements

I am indebted to many cattle breeders and breed
societies for their help in producing this book.

Picture Credits

(1) Kellythorpe Aberdeen Angus, (2) Ayrshire Cattle Society, (3) British Blue Belgian Cattle Society, (4) Temple Newsam Home Farm (Leeds City Council), (5) British Bazadaise Cattle Society, (6) British Blonde Society, (7) Angela Hamilton, (8) CIMARRON Brown Swiss, (9) The British Charolais Cattle Society, (10) Chillingham Wild Cattle Association, (11) The Devon Cattle Breeders Society, (12) Mark Bowles, (13) RE Archer Farms Ltd, (14) Helen McCann, (15) American Gelbvieh Association, (16) Gloucester Cattle Society, (17) English Guernsey Cattle Society, (18) Hereford, (19) Highland, (20) Holstein UK, (21) Irish Moiled Cattle Society, (22) National Milk Records, (23) The Kerry Cattle Society, (24) British Limousin Cattle Society, (25) Lincoln Red Cattle Society, (26) The Longhorn Cattle Society, (27) Luing Cattle Society, (28) Ben Beddoes, Dairy Dreams (29) Montbeliarde Cattle Society – Peter Bridge, (30) Murray Grey Beef Cattle Society, (31) The British Parthenais Cattle Society, (32) Craig Culley, (33) Wendy Robbins and Ian Stennett, (34) Callander Salers of Crieff, (35) Shetland Cattle Breeders Society, (36) Shorthorn Society, (37) British Simmental Cattle Society, (38) South Devon Herd Book Society, (39) The Sussex Cattle Society, (40) Temple Newsam Home Farm (Leeds City Council), (41) David Wynne-Finch, (42) www.rhuddelwelshblack.co.uk, (43) Temple Newsam Home Farm (Leeds City Council), (44) The Whitebred Shorthorn Society, (Aurochs) Porlock Visitor's Centre.

Foreword

Milk comes in bottles and cartons from the supermarket. Everybody knows that. When a group of children on a school trip were taken into a milking parlour they were amazed and shocked in equal proportions. Years of belief gone in a flash.

We have in the British Isles many different breeds of cattle, each with its own society or association. I have tried with their help to give you a glimpse of the tremendous diversity of cattle in these islands.

Some have been here for a thousand years while others are just off the boat, mainly from Europe. The purity of the Jersey breed is jealously guarded, unlike that of the black and white Holstein cow and its Friesian cousin; their complex relationship would need Sherlock Holmes to unravel.

Some breeds are as rare as the Giant Panda, others are more numerous — but all need our protection. In this book I have adopted the Rare Breeds Survival Trust classification to show the vulnerability of each breed. Many of the breeds in this book will become extinct if we do not support British producers by visiting farmers' markets and farm shops.

JACK BYARD
Bradford, 2008

Aberdeen Angus

Native to
The British Isles

Now found
On most continents

Protection category
 Op

Description

The Aberdeen Angus is predominantly black but red does occur. They are without horns, polled.

The Aberdeen Angus originated in North-East Scotland in the early 19th century and descends from the two local breeds of black cattle known as Hummlies and Doddies. Hugh Watson of Keillor in Angus is considered to be the originator of the breed. He bought quality stock from near and far then used only the finest polled black animals for his breeding stock. In 1842 'Old Jock', Watson's favourite bull, was born. Another star of the herd, a cow called 'Old Granny', was born in 1824 and is said to have lived for 35 years and given birth to 29 calves. Most of today's Aberdeen Angus can be traced back to these two animals.

The breed has a reputation for quality beef, established with the help of William McCombie. McCombie founded a herd based on Keillor stock and produced outstanding cattle which he showed in England and France. Development and improvement have continued into the 20th century.

Ayrshire

Native to
The county of Ayr
in Scotland

Now found
Throughout the British Isles
and on most continents

Description

The Ayrshire is any shade of red or brown with white. The patches are jagged at the edges and cover the entire body.

The improvement and development of the breed began in the mid 18th century when the native breed was crossed with Teeswater and Channel Island cattle. During this period, it was known as the Dunlop and then the Cunningham before becoming the Ayrshire.

By 1812 it was an established breed. For many years the horns were the hallmark of the breed. They were 30cm or more, curved upwards, outwards and backwards and when polished for the show ring were a magnificent sight. In modern farming, horns are impractical so today most Ayrshires are dehorned as calves.

The Ayrshire is a strong, healthy, long-lived animal and an effective grazer. This makes it capable of surviving in less than ideal conditions such as the heat of Africa and the extreme cold of Scandinavia, whilst still producing world-quality milk which is ideal for making yoghurt, cheese and ice cream.

3.

Belgian Blue

Native to
The British Isles

Now found
In Europe, Brazil, Canada,
New Zealand and the USA

Description

The Belgian Blue mainly comes in white, black and blue roan which is white hair over a base coat of a darker colour. Red is seen occasionally.

The Belgian Blue, as you would expect, has its origins in central and upper Belgium. During the latter part of the 1800s, Shorthorn bulls were exported from the British Isles to Belgium to improve the local red and black pied cattle. The early 20th century saw selective breeding to improve the quality of this dual-purpose animal with the major breakthrough in 1960. The modern Belgian Blue was the result of this skilful breeding.

The breed goes by many names: the Moyenne et Haute Belgique; Belgian Blue-White; the Belgian White and Blue Pied and the Belgian White Blue. In 2008 it was agreed that in the British Isles the breed would be re-named and promoted as the British Blue.

4.

Belted Galloway

Native to
Scotland

Now found
Throughout the British Isles, Australia, Canada, the USA and Switzerland

Description

The Belted Galloway is black, red or dun coloured with a white belt around the middle and is naturally without horns (polled).

The Belted Galloway's belted or 'sheeted' appearance creates more questions than answers about its origins though it is without doubt an ancient breed. The Belted Galloway, often affectionately called 'Belties', first appeared in Scotland during the 1500s although references to sheeted cattle can be found as early as the 11th century. The white belt is the only feature which separates the Galloway from the Belted Galloway. The characteristic was introduced by crossing the Galloway with the Dutch Lakenvelder.

Belties are double coated, having a soft short undercoat with a long shaggy overcoat that it sheds in hot weather. With about four thousand hairs per square inch the double coat is a good raincoat and ideal insulation against heat loss. In addition, for the health conscious, Belted Galloway meat has the same fat content as chicken and fish.

British Bazadaise

Native to
Bazas in France

Now found
In the British Isles, Australia, Belgium, Spain and Holland

Description

The cows are light grey and slightly shaded in wheat colour. The bulls are dark grey with a lighter saddle.

The Bazadaise (pronounced Baz-A-Day) originated in South-West France in the Middle Ages as a result of a cross between a local small grey cow called the Marini and a breed brought from Spain by the Moors. An exceptionally tough and vigorous animal, the Bazadaise will graze in the extreme cold of the high alpine meadows and the heat of the Spanish border region.

The breed was once valuable as a working animal because it is able to work in extreme conditions ranging from the lowlands to the Pyrenees. Farm mechanisation and war led to a decline in numbers.

In 1989 the Bazadaise was imported into the British Isles where the breed has steadily grown in popularity and is now a premier beef producer. In France the Bazadaise has achieved the coveted 'Label Rouge' which is used to indicate produce of a superior quality.

British Blonde

Native to
The Garonne Valley and
Pyrénées mountains in the
Aquitaine district of France

Now found
In the British Isles and
on most continents

Description

Predominantly a wheat colour but can range from almost white to brown.

Three strains of cattle have been used to develop the British Blonde: the Garonnais, the Quercy and the Blonde de Pyrénées. The Blonde has been grazing the pastures of Europe since the 6th century. They were originally used as draught animals and this continued until the end of World War II.

The breed, known in Europe as the Blonde d'Aquitaine, was 'improved' by various crosses with the Charolais, the Shorthorn or the Limousin but none were entirely successful and the Blonde was bred back to its original type. In 1974, the breed was introduced to the British Isles.

The British Blonde is strong, hardy, lean and docile and is renowned for the low fat quality of its meat that has led to its increase in popularity amongst the health conscious in Europe and the British Isles.

British White

Native to
The British Isles

Now found
In Australia and the USA

Protection category

Description

The British White is white with black or red points: eyelids, ears, feet, nose, and muzzle. They are polled (without horns).

The modern British White is a direct descendant of the feral white cattle of the British Isles. The original herd was mainly from Whalley Abbey in Lancashire and owned by Richard Assheton in 1553. Almost seventy years later his descendant Mary Assheton moved to Norfolk taking with her some polled white cattle from the original herd. This was the beginning of two major herds of British White cattle.

In 1865 Rinderpest or Cattle Plague arrived in the British Isles and 50,000 head of cattle had to be destroyed, leaving the British White at the point of extinction. The precautions taken at the time were the forerunner to those used 136 years later for Foot and Mouth Disease.

It was a constant battle to maintain numbers but work by the Rare Breeds Survival Trust and the British White Cattle Society ensured the survival of this beautiful animal.

Brown Swiss

Native to
The Canton of Schwyz,
Switzerland

Now found
In the British Isles, Europe,
the USA, South America,
Canada, Australia, South
Africa and New Zealand

Description

The Brown Swiss varies from a very pale brown to almost chocolate coloured with a creamy white muzzle, dark nose and dark blue eyes. When horned, the horns are short and white, growing darker toward the tips.

It is agreed that the Brown Swiss or Braunvieh are the oldest of all dairy cattle. This handsome breed originated in North-East Switzerland where skeletal remains similar to the Brown Swiss have been thought to date back to 4,000 BC.

Brown Swiss dairy cattle were developed from the Braunvieh dual-purpose cattle, the best milk producing Braunvieh having been chosen and selectively bred to create a quality milk-producing breed. Switzerland, the native home of the breed, has a reputation for producing quality cheeses.

In the summer months the cattle are moved to the mountains to graze on the sweet pastures that are a product of the heavy spring rainfall. The Brown Swiss is the second largest dairy breed in the world with over fourteen million animals.

Charolais

Native to
Charolles in South-East
France

Now found
In British Isles and
on most continents

Description

Charolais range from creamy white to tan with pink muzzles and pale hooves. There are also red or black varieties and they all have horns.

White cattle were grazing in the Charolles region in 878 AD and were known in the markets of Lyon and Villefranche. As with most Continental cattle, the Charolais was a multi-purpose beast so was used as a draught animal, for milk and for food. It was bred for utility not beauty. It took the French Revolution to give the breed widespread popularity.

In 1773 Claude Mathieu, a farmer in Charolles, moved with his white cattle to Nièvre in the region of Bourgogne where he improved the breed to such an extent that it became known, for a short while, as the Nivemas rather than the Charolais.

After the Second World War exports started to other parts of the world, first to Brazil followed by Argentina and South Africa. In the late 1950s the Charolais became the first Continental breed to be imported into the British Isles.

Chillingham Wild Cattle

Native to
The forests between the North Sea coast and the Clyde estuary

Now found
At the 300 acre Chillingham Park in Northumberland and another secret location.

Protection category

Description

The Chillingham Wild Cattle are white with long, curved horns and red ears.

The Chillingham herd was corralled in the late 13th century when the King of England gave permission for a wall to be built around the estate of Chillingham. The cattle captured ensured a supply of food and denied raiders the chance to rustle these cattle and drive them over the border.

These are not docile domestic cattle but feral, independent beasts; even in extremely harsh weather conditions they will not accept supplementary food other than hay. Their origins are lost in the mists of time though they are very similar to the Aurochs which are the ancestors of all modern cattle.

The catastrophic outbreak of Foot and Mouth disease in 1967 came within two miles of the Wild White Park so a small reserve herd was installed at a secret location as insurance against future disaster. After a further outbreak in 2001 genetic material was frozen to protect this magnificent herd against extinction.

Devon

Native to
Devon, Somerset,
Cornwall and Dorset

Now found
Throughout the British Isles,
Europe, the USA, Australia
and New Zealand

Description

Devon cattle are red, with the colour varying from a rich deep red to a lighter red.

The Devon is also called the North Devon to distinguish it from the South Devon which is a very separate breed. It is believed by some authorities that the Devon is descended from two breeds of indigenous British cattle called the Longifrons and the Urus as well as the native Devon cattle. There is also evidence that the Phoenicians brought red ancestral stock to these shores from North Africa or the Middle East on their frequent visits to buy tin. Could this be the reason the Devon is so adaptable to hot climates despite centuries of British weather?

The Romans also recorded the Devon Red Rubies in 55 BC. 1n 1623, 131 years after Columbus discovered America, the Devon was grazing the New World. The ship *Charity* delivered one bull and three heifers to the Plymouth Colony. They were the first pure-bred cattle to reach America.

Dexter

Native to
The south-west region of Ireland

Now found
In the British Isles and on most continents

Description

Predominantly black but also red or dun. Most Dexter cattle are horned but the polled is becoming more easily available.

It is frequently heard that the Dexter is a comparatively new breed. In fact, the breed has been around for over three hundred years. This dual-purpose breed was developed by a Mr Dexter who settled in County Tipperary in 1750. They are carefully selected from the hardy mountain breeds descended from the black cattle of the early Celts. The breed was further described in a report on Irish cattle in 1845 so is hardly a new kid on the block. The Dexter was introduced into the British Isles in 1882 by Mr. Martin of Oxfordshire and by 1892 was well established.

This is a small, gentle animal, 91cm to the top of its back (about the size of a St Bernard dog) and a specialist at browsing low-quality vegetation. The Dexter produces quality meat which is much sought after. In the early 20th century they were the show cattle of the gentry, but by 1970, the Dexter was rare and endangered. Their popularity with smallholders fortunately caused a dramatic increase in their numbers and saved them from extinction.

Friesian

Native to
Northern Holland
and Friesland

Now found
Throughout the world

Description

Predominantly black pied but red pied in small numbers.

The origins of the Friesian are unclear. Small black-and-white and red-and-white cattle were brought to Holland and Friesland from Jutland and crossing these with existing Dutch cattle is believed to have created the basis of the modern Friesian. The preference for the black pied led to breeding in favour of this colour though the red pied still exists in the Netherlands.

Friesians were imported into the east coast ports of the British Isles but this was halted in 1892 to prevent the import of Foot and Mouth Disease which was endemic on the Continent.

The Friesian is very closely linked to the Holstein although there is an increasing use of Friesian genetics because the benefits of the breed are being more appreciated. The society set up to support both breeds has changed its name several times to include or exclude the Friesian or Holstein names. In 1988 the Holstein name was re-instated and in 1990 The British Friesian Breeders club was established.

Galloway

Native to
South-West Scotland,
Dumfriesshire and Galloway

Now found
Throughout the British Isles,
Canada, Russia, the USA,
South Africa, Australia and
Alaska

Description

Predominantly black though the long outer coat may have a chestnut tinge and small number of red can be seen. The Galloway is polled.

The Galloway is one of the oldest and purest native cattle breeds. Records show that in the 16th century the native cattle of South-West Scotland were producing top-quality beef. The 1800s saw thousands of Galloway driven south to markets in Norfolk and Suffolk. The Galloway is an extremely hardy breed, capable of producing and calving in the harshest climates in places such as Russia and Alaska. They are aided in tough environments by their double coat which is a soft downy undercoat and long oily overcoat. The Galloway usually grazes wild upland countryside.

The Galloway is crossed with the Whitebred Shorthorn to produce the Blue-Grey. In 2001 Foot and Mouth Disease in Cumbria, Devon and South-West Scotland decimated the oldest and leading herds of breeding stock. Only the dedication and hard work of the breeders has returned the Galloway to the forefront of British cattle.

Gelbvieh

Native to
Bavaria in Southern Germany

Now found
In the British Isles, America and on most major continents

Description

The desired colour is a shade of yellow which one enthusiast describes as golden honey red, but black Gelbvieh are on the increase. They are usually polled.

The Gelbvieh (pronounced Gelp-fee and translated as 'yellow cattle') was developed in southern Germany at the turn of the 18th century and is one of the oldest German cattle breeds. The Gelbvieh was produced from local breeds in three districts of Bavaria to be used for food, milk and as a draught animal. Development continued into the late 19th century. In the mid 20th century red Danish cattle were introduced into the breed to improve milk production. The breed was imported into the British Isles in 1973 and has grown steadily in popularity.

Gelbvieh cows make excellent mothers and the calves are unusually small for such a large animal which results in fewer calving problems than similar European cattle. The Gelbvieh are noted for their quality milk.

Gloucester

Native to
The Cotswold Hills
and Severn Valley

Now found
Throughout the British Isles

Protection category

Description

The Gloucester is black-brown with a black head and legs. It has a white streak or 'finch' on its back and the white continues onto its tail and underneath. The skin on the muzzle and around the eyes is dark. The horns turn upwards and are fine and wide with black tips.

The Gloucester dates from the 13th century, making it one of the longest established breeds in the British Isles. This beautiful, rare animal was once a regular sight grazing in the fields of the Cotswold Hills and Severn Valley. The Gloucester is used for milk and beef and was once a draught animal. However, intensive farming methods, disease and the introduction of more commercial cattle led to the near extinction of this docile breed. In 1972 only one major herd remained in the British Isles and extinction was imminent. Only the dedication of five breeders ensured the survival of the breed.

The breed is known for Old Gloucester beef and Double Gloucester cheese and has gone down in history for its role in producing the Smallpox vaccine, thanks to a cow called Blossom.

Guernsey

Native to
The island of Guernsey
in the English Channel

Now found
In the British Isles, North
and South America,
Australasia and Africa

Description

The Guernsey is fawn and white.

There are many fanciful theories as to the beginnings of the Guernsey cow and the most common is unproven. In 960AD Robert, Duke of Normandy, sent a group of militant monks (the mind boggles) to defend the island against buccaneers and raiders, to cultivate the soil and educate the natives. They brought with them the finest French cattle, the Alderneys from the province of Isigny and the Froment du Léon from Brittany, with which they developed the Guernsey.

The Guernsey is renowned for the quality and the quantity of the milk it produces and the possible health benefits including protection against Type 1 diabetes, autism and possibly heart disease. The milk is richer in calcium than any other and is also high in Vitamin A and Beta Carotene which gives the milk its rich colour and are said to prevent a number of diseases including cystic fibrosis and arthritis.

Hereford

Native to
The British Isles

Now found
Throughout the world

Protection category
🐏🐏 🐏🐏 🐏🐏 🐏🐏 Op

Description

The Hereford has a dark red coat and a number of distinguishing white markings including face, crest, brisket, legs and tail. There are two strains of Hereford: horned and polled.

The origins of the Hereford go back to times immemorial. Records in Herefordshire show the breed as early as the 17th century where it began as a draught ox, pulling ploughs, carts and sleds. The Hereford is descended from small red cattle of Roman Britain crossed with a large Welsh breed that grazed the border of Wales and England. It took its name from the county in which it evolved.

In 1742, Benjamin Tomkins produced, with two cows and a bull calf from his father's estate, what is accepted as the beginning of the true Hereford breed. The early breeders created the superb beef qualities that are still apparent in the breed today. Herefords are the first English cattle to be recognised as a true breed.

Highland

Native to
The Highlands and Western islands of Scotland

Now found
Throughout the British Isles and on most major continents

Description

The Highland can be black, red, yellow, brindle or dun.

The Highland breed has grazed and browsed the Scottish Highlands and Western islands for over 1500 years and has played a major role in the area's development. There is speculation as to how they came to be there and the most likely suggestion is that they were taken there by the Vikings.

The term for a group of cattle is 'herd' unless you are of the Highland breed when you become part of a 'fold'. This term arose because the cowman would put his cattle into a fold at night to protect them from wolves and wild weather.

The Highland is double coated; a soft downy undercoat and a long oily overcoat which can reach 15cm in length. The outer coat protects against snow or rain and can be shed in hot weather. The long horns provide an excellent defence and the long forelock protects the eyes. For hundreds of years the Highland has provided food and drink for the small farmers or 'crofters' of the Scottish Highlands and islands.

Holstein

Native to
What is now the
Netherlands: the two
northern provinces of
Holland and Friesland

Now found
Throughout the world

Description

The Holstein has patterns of black and white.

The Holstein breed originated in Western Europe, particularly Holland and Freisland and is easily recognised as a large stylish animal. The breed as we know it began in the mid 19th century in the USA with the import of a bull and four heifers from Holland. The Holstein arrived in the British Isles at the end of the 19th century and in the years that followed approximately 2000 in-calf heifers were imported from the Netherlands along with several bulls.

After World War II a further 200 animals were imported from Canada which included three yearling bulls as a gift from Canadian breeders to help establish the breed after the war.

The Holstein will graze the fields during the summer months and are fed maize and silage made from grass during the winter. The average cow will produce approximately 9,000 litres of milk a year (15,800 pints). There is a complicated relationship between the Holstein and Friesian.

Irish Moiled

Native to
The north-west of Ireland

Now found
Throughout the
British Isles and Ireland

Protection category

Description

Predominantly red or roan with a white tail and under-parts and a white line (finching) on the back. They have dark eyebrows, a pink grey muzzle and are polled.

Originating in north-west Ireland this is one of the rarest cattle breeds in the British Isles. 'Moile' is Gaelic for 'little round' or 'mound' and refers to the rounded top of their heads. The 'Moily' is a very ancient breed with its origins possibly credited to the Viking invaders of the 8th and 9th centuries.

This traditional animal is valued by family farms in Ireland for its quality milk and beef production, achieved by allowing the Moilys to graze their natural food, grass. This technique is a slow and gentle process where quality of the product is more important than quantity. It is these fine qualities that have pulled the Irish Moiled back from the verge of extinction. In 1970 only twelve remained but now there are around a thousand which is a great improvement but a long way short of assured survival.

22

Jersey

Native to
British Isles and
North America

Now found
In the British Isles, Australia,
Canada, Denmark, New
Zealand, the USA and
Zimbabwe

Description

Most Jersey cattle are shades of fawn and cream although darker shades are common. They always have a black nose with an almost white border.

There is only one breed of cattle on Jersey and to maintain the integrity of the breed, cattle imports are banned and have been for 150 years. The Jersey is descended from the Guernsey and the breeds found on the Normandy and Brittany coasts; breeds that arrived in Europe from the Middle East.

The Jersey has been in the British Isles for 300 years and in its early days was called the Alderney. Her Majesty the Queen owns one of the oldest Jersey herds in the country at Windsor and there are hundred-year-old herds at Osberton in Nottinghamshire and Birthstone on the Isle of Wight.

Jersey milk is noted for its high quality rather than quantity since it is high in protein, minerals and trace elements. Its rich natural colour is derived from carotene extracted from grass.

Castle Learning Resource Centre

23

Kerry

Native to
Ireland

Now found
In the British Isles, Canada
and the USA

Description

The Kerry is black with black-tipped white horns, though some are de-horned as calves.

The Kerry is an Irish dairy breed and descendant of the Celtic Shorthorn which was brought to Ireland by Neolithic man as he migrated north from the Mediterranean. The Celtic Shorthorn is still found grazing the pastures of South-West Ireland as it has done for over 4000 years. The Kerry is acknowledged as being the oldest breed in Europe and the first breed developed primarily as a milk producer. The milk produced is of high quality and is ideal for cheese and yoghurt.

The Kerry is a manageable size, long-lived and hardy so can remain healthy even on poor quality grazing. The Kerry make ideal 'house cows'; one cow will rear several calves and provide sufficient milk for the average household. The Kerry became extinct on the British mainland in 1966 but was re-imported in the late 1970s. The breed is rare but growing in number thanks to excellent promotion.

24

Limousin

Native to
The Central Massif between Central and South-West France

Now found
Throughout the British Isles and on most continents

Description

The Limousin is mainly golden red, lighter under the stomach and around the eyes and muzzle. There are some black Limousin which are born light fawn or brown and darken with age. Their black coats are tinged with brown. The Limousin can be either polled or horned. The horns are fine and point forward.

Cave paintings in the Lascaux Caves near Montignac, France, depict cattle very similar to the Limousin which makes the breed an estimated 20,000 years old. The region in which the Limousin evolved is poor crop-growing country, so farming revolved around animals instead. The harsh, rainy conditions produced a sturdy, adaptable and healthy breed.

Records show that in the 17th century the Limousin was bred primarily as a draught animal and for food. In early 1971 the Limousin arrived in the British Isles. Leith Docks in Edinburgh received 179 cattle and within 15 years the breed became the main beef-producing animal in the British Isles, a record it holds to this day.

25

Lincoln Red

Native to
Lincolnshire

Now found
Throughout the British Isles
and on most continents

Protection category
😊😊😊 Op

Description

Deep cherry red in colour with white slightly incurving horns
though the trend is now for the Lincoln Red to be polled.

Records of the Lincoln Red go back to the late 17th century,
and DNA testing proves it was introduced into the British Isles
by the Viking invaders between 449 and 600AD.

In the late 18th century, the local Lincolnshire breed was
crossed with the Cherry Red Durhams and the York
Shorthorns, resulting in the Lincolnshire Shorthorn. In 1939
Eric Pentecost started to produce a breed without horns and
in 1960, the Lincoln Red Shorthorn became the Lincoln Red.

In the late 1970s, the breed's popularity suffered from the
importation of Continental breeds so it was successfully
crossed with selected European cattle to improve its
commercial standing. This was done with extreme care to
retain the Lincoln Red characteristics. The original Lincoln Red
is now under the umbrella of the Rare Breeds Survival Trust.

26

Longhorn (English)

Native to
North-West and Central England and Ireland

Now found
Throughout the British Isles

Description

The Longhorn comes in red, brown, grey, brindle coloured and varicoloured, but all have a white line down the back and large, wide, curved horns. Some are dehorned at birth.

The English Longhorn is a beautiful and ancient breed whose true origins are lost in time, although there are some prehistoric cave paintings depicting similar cattle.

In 1700 Robert Bakewell began developing the Longhorn with results which are still visible today. All animals are bred to suit contemporary needs and the Longhorn was developed when the horns were used to produce buttons and cutlery handles. When leaner meat was demanded and other materials were used for buttons, the Shorthorn took over.
By the 19th century the Longhorn's reign had ended.

The decline continued for 200 years and only the Rare Breeds Survival Trust prevented extinction. The Longhorn is now safe and produces quality healthy beef from eating its natural food, grass.

27

Luing

Native to
The Scottish Isle of Luing

Now found
Throughout the British Isles, Europe, Canada, New Zealand, Australia and South America

Description

Predominantly black pied but red pied in small numbers.

The Luing (pronounced Ling) originated on the Isle of Luing off the west coast of Scotland. The Luing is the result of skilful breeding by the Cadzow brothers in 1947 and produced from the cross of a Shorthorn bull and a Shorthorn-cross-Highland heifer.

The breed was developed through the economic need to produce a quality beef calf able to withstand the rigours of a harsh winter on the west coast of Scotland where it had to survive high rainfall and poor quality grazing. The Luing has a heavy winter coat so does not need as much to eat to stay warm. The winter coat is shed in summer.

In 1965 the British Government recognised the Luing as a breed in its own right and in the following years it was exported to all major continents. This strong, easily handled animal is exceptionally long-lived, usually reaching around twenty years. A beautiful animal.

28

Meuse Rhine Issel

Native to
The Netherlands and Germany

Now found
Throughout the British Isles, Australia and North America

Description

Red and white.

The Meuse Rhine Issel, usually called the MRI, was developed in the Netherlands during the 19th century on the banks of three rivers from where the name originates: the Maas, the Rhine and the Ijssel. In Germany the breed comes from the regions of Westfalia, Rhineland and Schleswig Holstein.

The German Breed Society was formed in 1900. From 1920 breeders in Denmark, Luxemburg, France and Belgium created their own red and white cattle using Dutch and German stock. Originally the MRI was bred for food and milk although many farmers have quite recently gone entirely into milk production. The milk has the protein variant Kappa Casein-B which is ideal for producing high quality cheese and perfect for making ice cream.

The MRI, which was first imported into the British Isles in 1970, is strong, docile and long lived. It has a high milk yield and produces top quality meat.

29

Montbeliarde

Native to
The Jura region of east central France

Now found
Throughout the British Isles

Description

Bright red and white. The white extends to the lower parts of the body, head, legs and tail.

The Montbeliarde cattle are descended from the Bernoise cattle brought to France by the Mennonites in the 18th century. It was originally called the Alsatian but in 1872 it was re-branded and entered the shows under the Montbeliarde name.

The breed is hardy and docile and adapts readily to different environments. It is renowned for its resistance to illness and its ability to survive in the harshest climates. The breed was first imported into the British Isles in April 1973 and numbers of purebred cattle have increased to several thousand.

The Montbeliarde is bred primarily for its milk but also produces top-quality meat. The milk is renowned for its high protein content and is traditionally used for making Emmental, Gruyère, Mont d'or, Bleu de Gex and Reblochon cheeses.

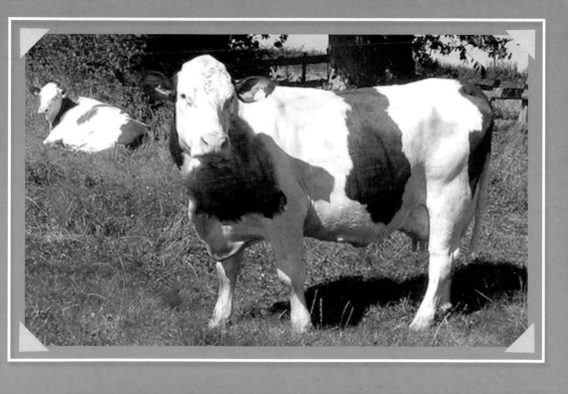

30

Murray Grey

Native to
The upper Murray Valley of Australia

Now found
Throughout Australia, the British Isles, New Zealand and North America.

Description

The Murray Grey has dark skin with a dark grey to light grey coat. It is polled.

The Murray Grey has progressed from an embarrassment to an international reputation. The first grey calves appeared by chance in a herd of black Aberdeen Angus cattle in the early 20th century on the Sutherland farm in Thologolong on the New South Wales border. The docile grey cattle grew quickly and were efficient at turning grass into meat. Local farmers soon took notice and began developing the breed.

By the mid 20th century the Murray Grey was a commercial success and the Murray Grey Beef Cattle Society was formed. The growth of the breed in Australia was unprecedented and its high quality was of the type that was in great demand in Japan and Korea.

The Murray Grey was first imported into the British Isles in 1970 and struggled because farmers tried to compare them to Continental cattle being imported at the same time.

31

Parthenais

Native to
The region of Parthenay
in the France

Now found
Throughout the British Isles

Description

From light tan to ginger-red with soft dark eyes and a black nose, hooves and ear tips. They are naturally horned but dehorned soon after birth.

The Parthenais (pronounced Par-Te-Na) is one of the oldest breeds in France and the herd book dates back to 1893. Until the 1950s the Parthenais was used as a draught animal for pulling ploughs and carts after which it became a beef breed but was also used for milk. It is a strong, adaptable animal with the ability to survive in extreme climates and all farming systems from intensive to ranch grazing.

The calves are dehorned shortly after birth to avoid damaging each other as they grow. The Parthenais was introduced into the British Isles around 1988 and has grown steadily in popularity since. For the health conscious, the Parthenais produces quality low-cholesterol lean meat.

32

Piemontese

Native to
The Piemonte region
of North-West Italy

Now found
Throughout the British Isles
and Europe

Description

The Piemontese is white or light grey. Newly born calves are golden brown but within months take on their adult colour. Older bulls become dark grey with darker patches on their head and neck.

The Piemontese is a truly ancient breed. Archaeological findings, fossil remains and rock paintings trace the breed's ancestors back to the Aurochs type. The Aurochs lived a minimum of 10,000 years ago which makes the Piemontese a truly ancient breed.

About 25,000-30,000 years ago another breed, the Zebu, migrated from what is now western Pakistan, reaching, but not crossing, the Alps. The Piemontese derives from the interbreeding of the Aurochs and Zebu. 'Piemonte' translates as 'at the foot of the valley', in this case the Alps. Rather later down the historical road, in 1988, the breed arrived in the British Isles. The breed is calm and friendly, a very desirable trait for farmers. For the health conscious, the meat has exceedingly low cholesterol content.

33

Red Poll

Native to
Suffolk

Now found
Throughout the British Isles,
Australia, New Zealand and
the USA

Description

The Red Poll is dark red or conker coloured with no white markings apart from the switch or 'tuft' of the tail. It is naturally polled.

The Red Poll is the result of crossing two ancient, and now extinct, breeds, the Norfolk Red and the Suffolk Dun. This was done in the early 19th century by James Reeve, a tenant farmer of the Holkham Estate. The Norfolk was the horned beef breed and the Suffolk a polled dairy breed with a history dating back to the Roman occupation of Britain. The result of the cross was a naturally polled dual-purpose breed producing milk and beef of superb quality. The Red Poll Cattle Society was formed in 1888.

The Red Poll is a hardy, long lived and economical breed which does not require large amounts of food to remain healthy. In the mid 20th century, Red Poll numbers went into a serious decline and it was classified as rare. The dedication of the breeders and the quality of the Red Poll has reversed this trend and the breed is once more taking its rightful place in commercial farming. 2008 saw the 200th anniversary of the breed.

34

Salers

Native to
The Cantel region of
south-central France

Now found
Throughout the British Isles

Description

The Salers is mainly dark red mahogany, but there are a growing number of black. Both can either be horned or polled.

The Salers (pronounced Sa-Lairs), is one of the oldest breeds in the world and has been browsing and grazing its homeland for over 7,000 years. Cave paintings near Salers, a small medieval town in France, depict cattle very similar to the Salers and the ancient Egyptian red cattle.

The Salers lives in an area with poor soil and a harsh climate, a mountainous region similar to the Lake District or the Scottish Highlands. It is a hardy and adaptable breed which is capable of producing beef and milk in difficult conditions such as cold winters and the occasional hot, dry summer.

The Salers, having been a draught animal and bred in the mountains, has strong legs, good feet and grows a thick curly coat in winter. In France only 10% of the herds are still milked and the milk is used to produce Cantal and Salers cheeses.

35

Shetland

Native to
The Shetland Islands

Now found
Throughout the British Isles

Protection category

Description

Predominantly black-and-white, but red-and-white occurs. They have short horns similar to those that you would expect to see on a Viking's helmet.

The Shetland is a truly ancient breed with remains having been found in archaeological digs which date back to the Bronze Age. It is thought to be a descendant of the Aurochs.

The Shetland was traditionally a house or crofter's cow and would supply a family with food and drink. Most crofters would have had two cows, one in milk and one in calf so as to ensure a regular milk supply. The Shetland is a calm animal which is easy to handle. It is a hardy animal which is able to survive on poor quality fodder; when the grass is gone the Shetland can eat seaweed and dried herring.

The modern Shetland is ideal for the smallholder because her lightweight frame makes her less likely to churn up good pasture in wet weather. For the health conscious, Shetland milk has good levels of unsaturated fatty acids and low levels of saturated fatty acids.

36

Shorthorn

Native to
The north-east of England

Now found
Throughout the British Isles
and most major continents

Description

The Shorthorn is red, red-and-white or white-and-roan. This particular roan colour is a mixture of red and white and is found in no other cattle breed. They are horned or polled.

The Shorthorn has evolved over 200 years from the Durham and Teeswater cattle of north-east England. In the late 1700s, the Colling brothers improved these two breeds using the techniques which Robert Bakewell developed.

In 1783, Charles Colling acquired four cows and at this time became aware of superior calves at a local market. These calves were bred from a bull called Hubback which Colling bought for £8. This shrewd move led to the birth of a bull named Comet in 1804 which Colling sold six years later for 1,000 guineas. This was the first recorded 1,000 guinea bull.

In the early 20th century the Shorthorn was a dual-purpose breed, but specialisation for milk and beef led to the breeders starting separate societies for milk and beef herds.

37

Simmental

Native to
Switzerland

Now found
Throughout the British Isles
and on most major
continents

Description

Colour varies from gold to red which may be evenly
distributed in defined patches on a white background. The
head is white and sometimes there is a white band across the
shoulders. They can be polled or with upturned horns.

The Simmental originated in the Simmen Valley in the Bernese
Oberland. It is now the most numerous breed in Europe and
the second largest breed in the world, exceeded only by the
Brahman.

The Simmental is a cross between a small Swiss native breed
and a large German breed and has a history dating back to
the Middle Ages. The breed was being exported to Italy as
early as the 15th century and to most of Eastern Europe, the
Balkans and Russia in the 19th. By 1895 it had reached South
Africa.

The Simmental is a docile, adaptable animal and is as happy
on a rural smallholding as it is in a more commercial farming
operation.

38

South Devon

Native to
An area of Devon known
as the South Hams

Now found
Throughout the British Isles,
Australia, New Zealand and
the USA

Description

A rich medium-red with copper tints, it can vary in shade.
Most South Devons are horned but polled animals do exist.

The South Devon, the largest of British native cattle, has
grazed in the south-west of the British Isles for over 400 years
and is thought to be descended from the red cattle imported
by Norman invaders in 1066. Because of their size and docile
nature the breed is known as 'The Gentle Giants'.

In 1620 the South Devon's predecessors were taken aboard
the *Mayflower* from Plymouth to the North American
Colonies where they were used to supply the staple needs of
the Royal Navy during the Napoleonic wars. Until well into
the 1800s this powerful animal had been relied upon to pull
the plough, as well as supply food and drink. In the 19th and
20th centuries careful selective breeding took place to further
improve the breed.

39

Sussex

Native to
The south-east of Britain

Now found
Throughout British Isles,
Australia, South Africa
and the USA

Description

The Sussex has a smooth dark red coat with white tail switches.

The Sussex is a truly ancient breed probably originating from the horned red cattle that grazed the dense forests of the Weald in Sussex and Kent around the time of the Norman conquest. The earliest mention of a pure-bred Sussex was in 1793 when Arthur Young gave the breed a glowing reference.

The Sussex was once noted for producing strong, powerful oxen ideal for working the heavy land, so was used as a draught animal until this work was taken over by horses and tractors. The Sussex was then bred for food.

The Sussex is a placid, adaptable animal with an amazing tolerance to heat and in the winter it grows a thick curly coat enabling it to survive the coldest Sussex conditions. It is also an efficient forager, able to survive and remain healthy on poor-quality grazing.

40

Vaynol

Native to
Vaynol Park in North Wales

Now found
As a protected herd
in Leeds

Protection category

Description

The Vaynol is predominantly white with black points: ears, muzzle, eye rims and feet, although all-black animals do occur.

The Vaynol is a unique breed of cattle and rarer than the Giant Panda. A semi-feral herd was established at Vaynol Park, North Wales and by 1886 numbers had risen to thirty-seven.

Vaynol parentage is mixed: white Highland bulls crossed with feral White Cattle. Further crosses were carried out between 1879 and 1886 using Highland, Ayrshire and Indian cattle. In 1980 the herd moved to Shugborough Park Farm when Vaynol Park was sold and in 1984 the herd was gifted to the Rare Breeds Survival Trust.

Two moves down the line and the herd of twelve animals arrived at Temple Newsam Home Farm in Leeds. By 2007 the herd had increased to thirty. Vaynols have a semi-feral nature and are aggressive to humans and cattle, including each other. The last surviving herd in the world can still be seen at Temple Newsam Home Farm Park where it is owned by Leeds City Council.

41

Wagyu

Native to
Japan

Now found
In the British Isles, Europe, the USA and Australia

Description

Wagyu are black or red though black is the dominant colour.

'Gyu' is the Japanese word for cattle and the term 'Wagyu' covers four main breeds. The Wagyu is considered to produce the caviar of beef. In 1976, several animals were exported to the USA, for research into improving American cattle with their superior genetic qualities. Prior to this, the export and breeding of Wagyu cattle outside Japan was forbidden.

Originally, the majority of the Japanese population were Buddhist and therefore vegetarian so the Wagyu was a draught animal used in the production of rice. However, the Shogun who ruled Japan from the late 12th century to the late 19th century found their warriors became stronger if they ate meat.

The Wagyu were given only the best grain to eat and beer to drink and were massaged three times daily. The latest research from Pennsylvania State University shows that eating Wagyu beef can reduce cholesterol.

42

Welsh Black

Native to
Wales

Now found
Throughout the British Isles
and on most continents

Description

Mainly varying from jet black to rusty black, but occasionally red. They can be horned or polled.

This native British breed has existed in the Welsh hills and mountains since long before the Roman invasion. It is a possible descendant of cattle from the Iberian Peninsula.

The modern-day Welsh Black is the result of ninety years of selectively breeding two Welsh breeds: the North Wales type, raised in the hilly and mountainous regions, and the South Wales type from a lower and gentler landscape. The Welsh Black is therefore a true British breed and possibly the purest breed in the world.

The breed is hardy and adaptable, growing a thick coat in winter which enables it to graze in snow and rain when most other breeds would head for cover. It is happy grazing in the lowland areas or foraging in the uplands. This adaptability enables the Black Welsh to survive when many other breeds would starve.

43

White Park

Native to
The British Isles

Now found
In the USA Germany,
Denmark, Australia and
Canada

Protection category

Description

Large and white with black points, ears, muzzle, eye rims and feet.
They have elegant, wide-spreading black-tipped horns.

The White Park is a rare and ancient breed that has been browsing
and grazing in the British Isles for over two thousand years. The
White Park must not be confused with the British or American
White because, though they share colour and looks, the White
Park is genetically a separate breed. Their nearest relatives are
the Highland and Galloway cattle of Scotland.

During the Middle Ages the landed gentry kept herds in enclosed
parks, but as fashions changed in the late 1800s the numbers
dwindled. The White Park was heading toward extinction. In 1941,
under threat of German invasion, five cows and one bull were
dispatched to Pennsylvania. Later additions were the beginnings
of the White Park cattle in America.

The White Park could possibly be the source of the word 'sirloin'
coined by James I. It is sometimes claimed that he enjoyed his steak
so much that he knighted it and so ate the very first 'Sir Loin'.

44

Whitebred Shorthorn

Native to
The England-Scotland border

Now found
As a protected breed

Protection category

Description

The Whitebred Shorthorn is creamy white.

The exact origin of the Whitebred Shorthorn is unclear but it created great interest in the England-Scotland border country over a hundred years ago when it was known as the Cumberland White.

Breeding of the Whitebred Shorthorn continued to grow until 1961 when, to preserve the purity of the breed, two hundred breeders formed the Whitebred Shorthorn Association. Cattle from 134 herds were inspected and 2310 females and 506 males created the first herd book. Despite sterling efforts from all the breeders the herd numbers began to decline, and in 2004 the breed was placed under the protection of the Rare Breeds Survival Trust.

The Whitebred Shorthorn is crossed with the Black Galloway to produce the Bluegrey, a very handsome animal. The breeds are valuable animals for conservation grazing as they reduce rank grasses such as Purple Moor Grass thereby opening up the areas to a greater number of species. They also produce superb quality beef.

Aurochs
(Bos primigenius)
Ancestor to most modern cattle

You will not see an Aurochs anywhere in the world today. It is the breed of cattle from which nearly all domestic cattle have descended and which became extinct in the British Isles about 3,500 years ago. According to records, it was black, stood 1.8 metres at the shoulder and had large curved horns. The last Aurochs died in Poland of old age, according to Royal inspection records, in 1627.

Castle Learning Resource Centre

RBST
Rare Breeds Survival Trust

The Watchlist covers sheep, cattle, pigs, goats, poultry, horses & ponies.
A breed whose numbers of registered breeding females are estimated by
the Rare Breeds Survival Trust to be below the category 6 "Mainstream"
threshold will be accepted into the appropriate Watchlist category.
In this book I have highlighted the first five categories.

Critical

Endangered

Vulnerable

At Risk

Minority

Further information: www.rbst.org.uk

In some cases categories refer only to the original population
of the breed. These are marked in the book with 'Op'.

Original photograph Bazadaise Society

Cow Talk

There are a few words related to cattle which to those not involved may seem like a foreign language. Here are just a few of these terms:

Cow	Female cattle
Bull	Un-castrated male cattle
Ox/Bullock	Castrated male cattle
Heifer	A young cow
Calf	Baby cattle before weaning
Brindled	A coloured pattern of spots or streaks
Pied	Patches of two different colours
Roan	A dark coat with spots or hairs of grey or white
Dun	A grayish-brown colour
Herd	The collective term for a group of cattle
Polled	A cow which is born without horns
Switch	The tuft at the end of a tail
Draught animal	An animal used for drawing heavy loads such as a plough or cart
Dual-purpose breed	One farmed for both milk and meat